Paula Andréia Bezerra Insaurralde
Juan Carlos Alvarado
Aiala Vieira Amorim

Using Ecological Stoves as a Sustainable Technology

Paula Andréia Bezerra Insaurralde
Juan Carlos Alvarado
Aiala Vieira Amorim

Using Ecological Stoves as a Sustainable Technology

A Case Study in the Garapa I Community in the Municipality of Acarape, Ceará, Brazil

ScienciaScripts

Cover image: www.ingimage.com

This book is a translation from the original published under ISBN 978-3-330-75707-3.

Publisher:
Sciencia Scripts
is a trademark of
Dodo Books Indian Ocean Ltd. and OmniScriptum S.R.L publishing group

120 High Road, East Finchley, London, N2 9ED, United Kingdom
Str. Armeneasca 28/1, office 1, Chisinau MD-2012, Republic of Moldova, Europe
Printed at: see last page
ISBN: 978-620-8-28535-7

SUMMARY

We dedicate this work to our mothers, who are the greatest examples of perseverance in the pursuit of knowledge and who, despite the difficulties, were able to pass on all their wisdom and constant support.

"A mind that is open to a new idea will never return to its original size.

Albert Einstein

THANKS

To God, who every day of my life has given me the strength to never give up.

To the University of the International Integration of Afro-Brazilian Lusophony - UNILAB, for the training.

To all the women farmers in the Garapa I community, especially Mrs. Makleny, for their availability and attention.

To my family and friends (old and new) who have always encouraged and supported me on this journey.

To my friend and coworker Neiliane Santiago Sombra Borges, for her support and companionship.

I would also like to thank Francisco Augusto de Souza Júnior (President) and José Tito Carneiro (Director) of ADAGRI - Agência de Defesa Agropecuâria do Estado do Cearà, for releasing me from my duties so that I could fulfill this dream.

SUMMARY

INSAURRALDE, Paula Andréia Bezerra. The use of ecological fires as a sustainable technology: A case study in the Garapa I Community, in the municipality of Acarape - CE. 2016. 51p. Master's dissertation - University of the International Integration of Afro-Brazilian Lusophony - UNILAB.

In recent years, society's concern about the environment and the use of sustainable technologies has increased considerably. In this context, various activities are being developed in the field of sustainable technologies in order to reduce environmental impacts, such as the proper management of solid and liquid waste, the use of renewable energy sources and local food production. The use of renewable energy sources has been intensified all over the world in an attempt to mitigate the harmful effects caused to the environment by the burning of fossil fuels. In the Northeast of Brazil, despite the widespread use of cooking gas, the use of firewood as a source of energy for domestic consumption in rural areas is still common practice. However, there are a number of alternatives for living in the semi-arid region with the aim of guaranteeing better living conditions for the population, including the use of ecological fires, which present an ecological proposal for reducing the consumption of firewood. Based on this information, the aim of this project was to analyze the impact of ecological fires on the quality of life of the beneficiaries and on the preservation of the environment in the Garapa I community, located in the municipality of Acarape, Cearà. To this end, a case study was carried out by applying questionnaires to the women who benefited from the ecological fires, as well as measuring the temperature of these fires. According to the analysis of the general data from the survey carried out with the users of the ecological stoves, it was found that the objectives that the stoves reduce the consumption of firewood and eliminate smoke inside the home were not achieved, since the majority of the interviewees said that they felt no difference in their health when using the ecological stove. The use of this technology has a negative effect on the health of the interviewees, especially their breathing and the high temperature of the environment. The beneficial effects on health were not properly perceived in ecological stoves, as the smoke remains inside the house. The importance of disseminating improved wood-burning technologies in rural areas is undeniable, but first it is necessary to prove that these technologies will actually meet their objectives.

Keywords: sustainability, semi-arid region, firewood, energy matrix.

1 INTRODUCTION

In developed and developing countries, the use of firewood by communities is common, especially in rural areas. This is because, since the discovery of fire, its use has been of great importance to humanity. Its importance is due to its use for cooking (KAMIMURA; BURANE, 2010). With this, its application reinforces the influence that this type of material has over other existing sources of energy.

The wood-burning stove is one of the oldest and most widespread technologies for providing energy. According to Simon, this technology is used by 3 billion people (SIMON, 2011; UNDP 2009). In rural areas, its use is due to its low cost compared to the cooking gas used in conventional fires.

In Brazil, the use of firewood for domestic consumption in rural areas is still common practice, despite the widespread use of cooking gas. In the residential sector, firewood is the second largest energy source, behind electricity and liquefied petroleum gas (LPG), which are in third place (BEN/MME, 2011).

The emission of smoke and particulates into the internal environment of homes is one of the major problems with the use of wood-burning stoves, in addition to the low energy efficiency resulting from the incomplete combustion of wood, which in the long term can cause serious health problems for users (SOUZA *et al.*, 2003).

Currently, in developing countries, the use of biomass fires is the fourth leading cause of death, and women and children are the groups most affected, as they are more exposed to the problems caused by the combustion of wood fires. It is estimated that the number of people who could be affected by occupational and environmental health problems due to the use of biomass combustion inside homes could exceed 200 million by 2030 (SOUZA *et al.*, 2003). With the high price of LPG, the consumption of firewood in rural homes is increasing, leading to the contamination of this population by the smoke expelled by fires.

Wood-burning fires are quite common in rural dwellings, especially in Latin America, where traditional buildings predominate and in the vicinity of these buildings there is a potential for the development of agroforestry agriculture and a potential for firewood (MAIA *et al.*, 2008).

Although the impacts on human health, the environment and social development related to the rudimentary and inadequate use of wood-burning fires are well known in Brazil, few efforts have been made towards the clean use of wood in homes (WINTOCK, 2007).

Faced with all the problems caused by the use of wood-burning fires, in 2005 the NGO IDER - Institute for Sustainable Development and Renewable Energies, began to set up ecological fires in rural areas of the state of Ceará, with the aim of remedying the main problems caused by traditional fires, which are: poor energy efficiency, causing greater consumption of wood and smoke inhalation, especially by women and children, who spend the most time near the fires.

Based on the above, the aim of this book was to analyze the impacts of ecological fires on the quality of life of the beneficiaries and on the preservation of the environment in the Garapa I community, located in the municipality of Acarape, Cearà.

2. OBJECTIVES

General objective

To analyze the impact of ecological fires on the quality of life of the beneficiaries and on the preservation of the environment in the Garapa I community, located in the municipality of Acarape, Ceará.

Specific objectives

- Carry out a preliminary study of the impacts of burning wood on ecological fires and on the quality of life of the users of these fires.

- Analyze the temperature of the kitchen when the equipment is in operation.

3. LITERATURE REVIEW

3.1 The energy and environmental crisis

In the process of generating energy, man has several alternative ways of obtaining energy, ranging from hydroelectric power to biofuels and even nuclear energy (KATWAL; SONI, 2003). According to data from the Energy Research Company (EPE, ANO 2012), the main source of energy used in the residential sector is electricity, followed by liquefied petroleum gas (LPG), firewood, natural gas and charcoal. As the supply of these energies involves the use of natural resources, it is of the utmost importance that these are exploited carefully so that atmospheric pollution and soil degradation do not occur. Faced with this difficulty, society needs to understand what resources are available in nature and how they should be used in order to reduce their environmental impact.

Facts show that existing natural resources account for 50% of the productivity used by human beings (VITOUSEK *et al.*, 1997). Man has made use of these resources, but he needs to worry about future generations, because data from the United Nations Organization indicate that by 2030, planet Earth will face a water shortage that could affect 40% of the world's population. In addition to water, another resource used as an energy source is oil, which is considered to be the energy matrix that drives the world. The use of this non-renewable resource as an energy source has been questioned, generating debates involving political and environmental issues (REIS; SILVEIRA, 2000). According to Bermann (2008), it is in the world's interest that dependence on fossil fuels should be replaced by clean, renewable and environmentally friendly energy.

In the midst of this situation, there is concern about the damage caused to the environment. In this way, technologies are being sought that can cause less environmental impact in energy production, as well as a more sustainable form, which, according to the Brundtland Report (1987), consists of meeting the needs of the present generation without affecting the ability of future generations to meet their needs. In the case of Brazil, this has been highlighted in international forums, with the aim of improving the use of renewable energies as well as diversifying energy generation sources. Such a commitment reduces the risk of a new hydrological deficit, which usually leads to a crisis and rationing, as happened in the Brazilian summers of 2001/2002, as well as the one we are currently experiencing.

In this context, the environmental crisis has brought back the discussion of the relationship between man and nature. In reality, it is due to actions taken without any concern for the preservation and/or maintenance of natural resources. In recent years, society's concern for the environment and the use of sustainable technologies has increased considerably. Nowadays, companies are required to be more careful about pollution, deforestation, global warming, water shortages, the semi-arid region, etc (ALCÓCER *et al.*, 2015).

The environmental crisis goes beyond local barriers and affects the whole of humanity, causing an unfavorable socio-environmental situation for the quality of life of human beings. The risks of this crisis require human action not only in understanding the problems related to the environment, but above all, a change in attitude (ANDRADE, *et al.*, 2012). It is worth noting that the environmental crisis not only affects the natural environment, but also has a socio-economic impact that jeopardizes the sustainable development of the regions affected by climate change and the reduction of water resources.

The emerging economic crisis and environmental degradation have stimulated the emergence of sustainable technologies, the aim of which is to prevent the collapse of the ecological system and the form of disorderly development rooted in the cultural, social and political spheres of today's society (INSAURRALDE *et al.,* 2015). Sustainable development isn't just about saving natural resources, it's about reflecting and thinking about something broader, that has an economic, social and environmental character, something that can be left to future generations.

In order to minimize environmental impacts, various activities are being developed in the field of sustainable technologies, such as the proper management of solid and liquid waste, the use of renewable energy sources and local food production. These techniques, when it comes to sustainability, contribute to a better performance of the social, economic and environmental indices of populations (SATTLER, 2003). Initiatives with this objective are necessary to preserve natural resources and contribute to improving people's quality of life.

With regard to the production of renewable energy, various proposals have been put forward, including wind energy and biomass, which uses sugar cane to produce ethanol. As a way of encouraging the use of these energy sources, the Incentive Program for Alternative Sources of Electricity (PROINFA) was created in Brazil on April 26, 2002, by Law No. 10.438, with the aim of getting municipalities to participate in an incisive way (BERMANN, 2008). There are also proposals that do not present a new energy matrix, but provide a lower negative environmental impact, consequently greater conservation of nature, presenting possibilities for reducing the amount of use of natural resources. Among the technological proposals, there are large-scale ones, usually for industrial production, such as biodiesel, and the so-called social ones, which are cheap, easy to build and access, such as biodigesters and ecological fires.

3.2 Sustainable technologies and coexistence with the semi-arid region

Coexistence means living together with others, "with living", being together. In this way, coexistence with the semi-arid region implies a harmonious relationship between people and the environment, because it is not a question of trying to change the natural characteristics, but of adapting to them. For there to be a balance between the environment and the various living beings, there needs to be interaction and acceptance between them. "[...] acceptance of and care for the other, recognized in their legitimacy as a shared other, the one with whom each of the parties to the coexistence establishes ties of complementarity and interdependence" (PIMENTEL, 2002).

In order for this coexistence to be harmonious, it is necessary to adopt sustainable adaptation practices, such as improving water collection and storage (Figure 1), as well as rainwater management and the sustainable management of water sources. To this end, it is necessary to adopt alternative technologies for capturing and storing this water, thus enabling the development and growth of family farming and consequently improving people's living conditions, as well as combining appropriate crops for each region and the sustainable management of native vegetation.

Figure 1. Photograph of a slab cistern used to store rainwater.

Source: deolhonocariri.com.br

With the aim of applying a new development model aimed at the social inclusion of the most economically fragile populations, the use of sustainable technologies has begun to spread (ALCÓCER, *et al.*, 2014; ALCÓCER, *et al.*, 2015). For this to happen, it is essential that local people take an active part in the development of family farming, where they can raise awareness and prioritize the search for solutions that are appropriate to the local natural conditions. It's no good just owning the land and the technology to use it, but respecting and caring for it is necessary to avoid failures and inappropriate use of the soil, which can lead to its degradation.

3.3 Ecological fires as sustainable technologies

Ecological or eco-efficient fires are proposals presented as ecological, to reduce the consumption of the vegetable energy matrix, in this case firewood, and to coexist with the semi-arid region. They are social technologies built as alternative models to traditional wood-burning fires, which are still widely used in rural communities. These fires were developed with the aim of generating more energy by burning less wood, using only sticks and other materials, thus contributing to reducing deforestation in the caatinga biome (MMA, 2013).

Another important aspect is that, in addition to conserving the caatinga biome, there is a prospect of improving the quality of life of the people who use these fires. This benefit is expected because the stoves are sealed, giving off all the smoke through the chimney, which usually pollutes the whole house, especially the kitchen. Some of this smoke was inhaled mainly by women and children. Some research has shown a reduction in the rate of respiratory and eye diseases among the members of the families where the stoves were built (SANGA, 2004).

This technology, called social technology, is considered to be accessible and democratic, as it allows the beneficiary families to manage the equipment. According to the Social Technology Network, the concept of social technology "(...) includes products, techniques and/or methodologies that can be applied, developed in interaction with the community and which represent effective solutions for social transformation". In addition, in the quest to solve social problems, ecological fires also meet the requirements of simplicity in handling, low cost, easy applicability and proven social impact, contributing to social inclusion, generating work and income and promoting sustainable local development.

In Brazil, the use of these fires has been carried out mainly in the northeastern states of Bahia, Cearà, Pernambuco, Alagoas and Minas Gerais (CARVALHO, 2012). In Cearà, between 2007 and 2011, around 26,500 ecological fires were developed and distributed by the Institute for Sustainable Development and Renewable Energies (IDER). According to data from the Ceará State Government's Department of Cities, the ecological fireplace project began in 2007 as part of the Housing Improvements program. R$8,798,681.49 had been invested by 2014. During this period, 17,000 families benefited from the delivery of ecological fireplace units, 13,497 of them using the previous masonry model and 3,500 using the current model, which is more energy efficient.

The ecological stove consists of a metal base, a chimney and refractory bricks that concentrate heat in the three burners and the top plate. The metal structure is made up of a base frame, top plate, inner plate, grate, drawer, lid, burner plate and three burners. All these structures are made of iron, and the plates and the mouth must be cast iron. The base is made of bricks, with a minimum height of 65 cm, a length of 130 cm and a width of 60 cm. The masonry is made of bricks and refractory mass. It also has a chimney that concentrates the heat in the burners and the plate and exhales all the smoke (Figure

2). In this way, the ecological stove guarantees a much more efficient burning of biomass, thus reducing the consumption of firewood and guaranteeing the preservation of the environment.

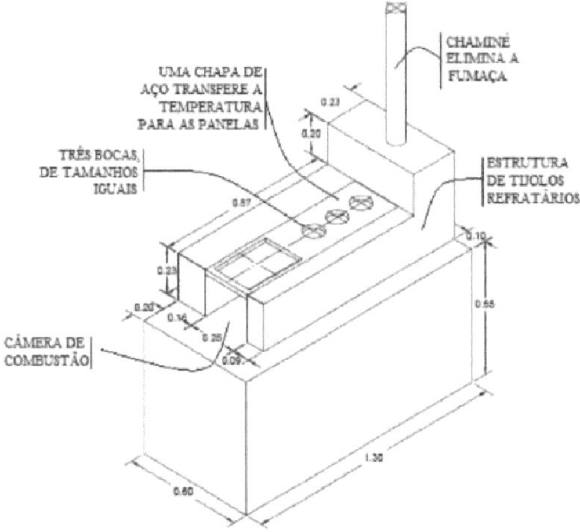

Figure 2 - Schematic model of an ecological stove. Source: author

Due to the good acceptance of the Ceará beneficiaries, the state government decided to turn this idea into a public policy, benefiting low-income families living in rural Ceará who still depend on wood-burning fires (Figure 3). The new fires, implemented by the secretariat for cities, reduce the environmental impact by making better use of the heat generated by burning wood.

Figure 3: Photograph of Mrs. Makleny, a resident of the Garapa I community, lighting the ecological stove.

Source: author

The process of disseminating this technology began with a course on building, operating and maintaining the technology. A mobilization period was defined for each course given, with the aim of increasing the organization and participation of the settlers. In the first course, a strategy was adopted to mobilize the community, covering different localities, in order to pass on knowledge of the technology and identify multiplier agents. However, due to logistical problems, it was later decided to concentrate the actions of each course in a single community (Institute for Sustainable Development and Renewable Energies - IDER).

Each ecological stove also has a chimney, completely eliminating smoke from the home environment, thus helping to reduce cardio-respiratory diseases, which mainly affect women and children. The project also helps to generate income in the communities involved, as the fires are installed by bricklayers from the towns themselves. In addition, before the fires are installed, community meetings are held to discuss with the families the importance of preserving the environment and valuing the caatinga (IDER).

In this way, ecological fires have emerged to oppose the conventional wood-burning models used in rural areas, which have low energy efficiency because they make incomplete use of the wood, generating the emission of polluting gases and particles into the environment around the stove and into the atmosphere. As well as emitting smoke into homes, these conventional fires also consume more wood, since the heat is incompletely utilized and burning occurs more quickly (BORGES, 1994). It is worth noting that the emission of polluting gases and particles from conventional wood and coal fires generates pollution that causes diseases such as respiratory failure, eye diseases, chronic bronchitis and others. These respiratory diseases cause the death of thousands of people every year.

3.4 The Brazilian Northeast and the use of firewood as an energy source

The Northeast region of Brazil represents 18.3% of the national territory and is made up of nine states. According to IBGE estimates (2015), the Northeast region has a population of 56.5 million inhabitants. The so-called semi-arid region has approximately 26.5 million inhabitants, and its characteristics and environmental conditions are very severe. According to Silva (2006), the semi-arid region is characterized by low rainfall, soils poor in organic matter and regions with an arid climate, which characterizes seasonal aridity.

The characteristics of the northeastern semi-arid region, on the occasion of the caatinga biome, are a dry and hot climate, with rainfall that is poorly distributed in both space and time, and which is centered on the summer and autumn seasons, these rains occurring at short intervals (LEAL *et al.*, 2003, p. 75). Rainfall in this region can vary from 500 to 800 mm, which is very little compared to the southern regions of the country, but compared to other semi-arid regions in the world, it is one of the wettest with an average annual rainfall of 750 mm. These adversities cause serious problems in the production process of the people who live in this region, especially for family farmers.

According to Vesentini (2000), the Brazilian semi-arid region has its own particularities that are already well known and studied, such as drought, which is a natural problem that must be tackled in order to find ways of living with it that are adapted to its characteristics. For this significant portion of the national territory, there is a lack of efficient public policies that deal with the issue of periodic droughts on a permanent basis, with the effective participation of the people who experience their reality.

All these situations end up interfering with living conditions and well-being, especially for the population living in rural areas. Government authorities have been paying a little more attention to this problem in recent years, as well as arousing the interest of non-governmental organizations (ZANELLA, 2014). Thus, various programs aimed at coexistence with the semi-arid region have been and are being developed and implemented, seeking solutions that improve and facilitate the living conditions of these people, as well as subsidizing the definition of public policies for this region.

The first energy source used by humans to make fire was firewood. Later, firewood was used to heat, cook food and light the environment, as well as being used as a weapon to hunt and defend against ferocious animals. It is very important in Brazil's Energy Matrix, accounting for around 10% of primary production, and can be obtained through plant extraction from reforested regions or native forest (BITTENCOURT, 2005).

Its use is due to the fact that firewood is a low-cost energy source, which has led to it being called the energy of the poor, since in developing countries it accounts for about 95% of the energy source, while in industrialized countries, the contribution of firewood reaches a maximum of 4%. The use of firewood has been one of the causes of deforestation in the semi-arid region of Brazil. Only 20% of the firewood consumed in the region comes from sustainable management (UHLIG, 2008).

In Brazil, approximately 40% of the firewood produced is transformed into charcoal, 29% of which is consumed in the residential sector, followed by the industrial sector with approximately 23% of consumption (USP, 2016). The main wood-consuming industries are the beverage and food industries, as well as the paper and ceramics industries. As for residential consumption, firewood is mainly used for cooking. A family of 8 people needs approximately 2 m^3 of firewood to prepare their meals for a month (BITTENCOURT, 2005).

Firewood has always been an important source of energy for humanity, but with the technological advances that have taken place in developing societies, it has lost its place to fossil fuels. As a renewable energy source, biomass is one of the energy sources that offers the best prospects for the future (BITTENCOURT, 2005). The following are considered renewable energy sources: solar, wind, biofuels, biomass and others. According to Bacchi (2006), worldwide renewable energy represents 20% of total energy consumption, with 14% coming from biomass.

There are two concepts related to biomass as a renewable energy source: renewal and sustainability. Renewability is a characteristic of the energy source, while sustainability is the way in which this source is managed (UHLIG, 2008). According to the author, wood biomass is among what can be considered renewable energy. Sustainability occurs when nature is replenished faster than the rate of use. For this to happen, management needs to be done in a way that is compatible with replenishment capacity, so the sustainability of some renewable energy sources depends on proper management.

Looking at the use of firewood in the Brazilian Northeast, in the semi-arid region, forest products are predominantly used for the production of firewood and charcoal. In rural households, these energy sources are mainly used for cooking food, and in the commercial and industrial sectors, the use of forest products is related to the drying and burning stages in the production process (MMA, 2013). Due to the use of firewood as an energy source and the replacement of these native species with crops and pastures, the caatinga biome is in a high degree of degradation.

Among the main native species of the Caatinga that are used for firewood are angico (*Anadenanthera macrocarpa*), angico de calro (*Piptadenia obliqua* (Pres.) Macbr.), catingueira rasteira (*Caesalpinia microphyla*), sete-cascas (*Tabebuia spongiosa*), aroeira (*Myracrodruon urundeuva* Engl.), baraûna (*Schinopsis brasiliensis* Engl.), jurema preta (*Mimosa tenuiflora* (Willd.) Poiret), pau d'arco (*Tabebuia impetiginosa* (Mart. ex DC.) Standl.), the creeping true catingueira (*Caesalpinia pyramidalis* 6 Tul.), sabià (*Mimosa caesalpiniifolia* Benth.) and umburana (*Commiphora leptophloeos* Engl.) (DRUMOND *et al.*, 2000).

Despite the fact that the conditions for exploiting its resources in the medium and long term are not guaranteed, the caatinga has been the main source of wood for both residential and industrial fuel (BRITO *et al.,* 1997). This vegetation accounts for between 30% and 50% of the primary energy in the Northeast. In rural areas, it is a widely used source of energy. Approximately 80% of the wood harvested from the caatinga is consumed as a source of energy, and its consumption as energy is the sector that generates the greatest extractive pressure in the Northeast (CAMPELLO *et al.,* 1999). The degradation of the vegetation as a result of various inadequate agricultural practices has contributed to increasing the levels of desertification in this region.

The native forest, which once seemed inexhaustible, has always been a source of firewood. But the devastating way in which it has been exploited has left our country in a critical situation: regions where there used to be large forest cover are now in the process of desertification, soil degradation and altered rainfall patterns, all due to deforestation. Biomass used for combustion, especially firewood and coal, is one of the oldest energy alternatives used by humanity to cook food (MORAES; MARTINS; TRIGOSO, 2007). In developing countries, the majority of the population living in rural areas uses biomass, especially firewood, as the main and most accessible source of domestic fuel for cooking, heating water and the interior of homes.

The use of renewable energy sources has been intensified by developed and developing countries in an attempt to mitigate the harmful effects caused to the environment by the burning of fossil fuels (PARIKKA, 2004). This intensification of renewable energy sources is also due to their natural replenishment capacity, while non-renewable sources could, with intensive use, be depleted.

Since the 1980s, especially in developing countries, government and non-government programs have been set up to develop and disseminate cleaner-burning and more efficient fires (SANGA, 2004). In Brazil, the study and first clay prototype of this technology was developed in 1992 in the communities of Mocambo and Caburi, in the municipality of Parantins/AM (MARTINS et al., 1992). In Ceará, it was the Institute for Sustainable Development and Renewable Energies (IDER) which, in 2005, experimentally set up twenty units in the municipality of Itapipoca (CE).

4. RESEARCH METHODOLOGY

4.1 Theoretical background

The theoretical basis for the analysis presented in this work was based on a bibliographical review of the subject. The research was based mainly on reports from Non-Governmental Organizations (NGOs), research on the websites of the Ceará State Government and the IBGE, as well as research in papers, articles and books on the subject.

4.2 Description of the study site

As a specific case study for this research, a small low-income rural community was selected, located in the semi-arid region of the Baturité Massif in the municipality of Acarape. This municipality is 52.2 km from Fortaleza, the capital of Ceará. Its name comes from the Tupi "acarà" and means from acara pé, the path of the acarâs, the fish channel or the path of the herons. Before 1926, its name was Calaboca. Acarape has an approximate area of 155.188 km^2 , with 16,288 inhabitants and a population density of 98.52 inhabitants km^- 2 (IBGE, 2010). Its altitude is 95 meters, its Human Development Index (HDI) is an average of 0.606 (IBGE, 2010) and its Gross Domestic Product (GDP) *per capita* is R$ 6,260.49.

The municipality of Acarape's main sources of water are the Acapape River and the Pacoti River, which cut through its territory, and several reservoirs such as Hipólito and Boqueirâo. Its climate is tropical hot semi-arid with rainfall concentrated between January and April, with an average rainfall of 1,097 mm. Located between the mountains of the Maciço de Baturité and the Serra do Cantagalo and on the banks of the Acarape-Pacoti river, Acarape was inhabited by the Potyguara, Jenipado, Kanyndé, Chorò and Queito ethnic groups, and received several military and religious expeditions from the 17th century onwards. The town of Calaboca, part of the village of Redençâo, arose from the development of the region, especially from the 18th century, with the production of sugar cane, the consequent creation of stills and mills in the region and the arrival of African slaves to carry out this work.

The community in Acarape chosen for this research is called Garapa I. The Garapa I community is located 12 km from the center of the city of Acarape and is a region where firewood is available for use in cooking fires, which was one of the relevant criteria in choosing the case study. In addition, this was a location mentioned in several reports on ecological fires, as a reference and one of the first regions where the fires were built and tested. Another contributing factor was the contact with the local leader, through her work on the municipal committee of FCSVA - Fòrum Cearense pela Vida no Semiàrido (Ceará Forum for Life in the Semi-Arid), contributing to the process of building plate cisterns in the municipality in support of the work carried out by Espiar - Centro de Pesquisa e Assessoria (Research and Advisory Center).

According to IDER (2012), 30 new ecological fires have been installed in the Garapa I community, benefiting 50 families. Access to this community is via gravel roads by car or motorcycle. For the community, there is a so-called hourly car that travels to the municipal seat, leaving early and returning at 12 noon. At any other time, you have to pay between 10 and 12 reais to rent a motorcycle and 30 to 40 reais for a car.

As far as organization is concerned, Garapa I is organized as an agrovillage, with houses close together. There is an evangelical church, a building housing the community association, a small shop and a building housing the old school, which is currently inactive due to the risk of collapse. The community's children are therefore studying in two neighboring communities, Garapa II, 3 kilometers away from Garapa I, and Amargoso, 10 kilometers from Garapa I. The school is transported by bus.

As far as health care is concerned, the community only receives a visit from the PSF - Programa Saù da Familia (Family Health Program) once a month, when they are attended by a doctor and a nurse. Each day, 17 cards are distributed to each professional, with the nurse giving priority to hypertensive patients, diabetics and pregnant women (prenatal care). Attendance used to take place at the school, before its structure became unsafe, and now takes place at a resident's house.

Purchases for family consumption are made at 3 larger shops in Acarape or in Redençao, a neighboring municipality. A large number of Acarape residents go to the fruit and vegetable market in Redençao on Sundays, where they take advantage of the opportunity to shop for their general needs.

Access to water is via slab cisterns for human consumption. There is no piped water in the community, and pipes were installed a few years ago, but they are still not receiving water. The project would bring water from the Acarape do Meio reservoir, a neighboring town. For domestic use, bathing, washing dishes and clothes, families take water from the pond in the community when it is full. There is still a strong habit of women washing clothes in the lagoon, but they take care to remove the water with buckets so that they can use it in the downhill stone beaters. This is also the water used for animal consumption. During periods of drought, the lagoon dries up, making access to and consumption of water extremely difficult.

The main productive activity of the families surveyed is family farming, with the biggest crops being corn, beans and fava beans. We have seen that deforestation and burning are still used to clear production areas and pesticides are used to kill so-called pests. They also raise small animals such as chickens, pigs and goats. Many even keep donkeys, which are used as pack animals to carry water and firewood. We know that in the backyards, women in particular also grow vegetables, medicinal plants and fruit trees.

Most of the people interviewed still use the wood stove, in addition to the gas stove. The firewood is collected from nearby forests. Even though we know that families use firewood on a daily basis, we were struck by the fact that two families in the community make their living from selling firewood. They pay for the use of the forests, cut the wood and, in "good" times, take up to 6 cartloads/trucks of firewood a day to neighboring towns. (Information provided by resident Antônia Makleny de Sousa Pastor).

4.3 Type of study

A case study was carried out by administering questionnaires to the families who had benefited from the ecological fireplaces, with the aim of surveying the perception of the users of these fireplaces about the impacts on their lives and internal environmental comfort. A survey was carried out on the current state of use of ecological fireplaces and possible environmental impacts. Questions were also asked about the impact of the use of ecological fires on the respiratory health of users.

The interviews and field study took place between August 2015 and May 2016, with the collaboration of the community leader of Garapa I, Mrs. Antônia Makleny de Sousa Pastor, as well as the beneficiaries of the ecological fires. In addition to the interviews, ambient temperature measurements were taken before and during the use of the fires, using a digital infrared thermometer with a laser sight (-50° to 380° C), model: DT8280, specifically for this purpose. Temperature measurements were also taken at various points in the structure of the ecological fireplaces, before, during and after use, using a pistol-type thermometer (Figure 4). This type of thermometer is used to measure the surface temperature of an object and can be used on hot, dangerous or difficult to reach parts without safe and quick contact.

Figure 4 Measuring the temperature of the ecological stove before use, in the Garapa - I community, in Acarape - CE.

Source: Author

4.4 Organizing the results

The interviews were transcribed in full (see the script used in the interviews) and then the data was tabulated and analyzed. When drawing up the questionnaires, care was taken to avoid questions that could give rise to arbitrary, ambiguous or unfocused, absurd or empty statements, which would detract from the meaning of the interview. A logic of thought was also sought, directing the content towards a continuity of themes, avoiding the 'back and forth'. The interviews were carried out with the beneficiaries of the ecological fires in the garapa I Community, on previously scheduled dates and times, always with the help of Mrs. Makleny (community leader), who was responsible for mobilizing the interviewees. After transcribing the interviews, they were reviewed and 25 (twenty-five) beneficiaries were interviewed.

To carry out the preliminary analysis of the data, tables were formatted using Word software and then spreadsheets and graphs using Excel software, highlighting the main topics of the interviews and the most essential elements.

According to Machado (2003), when the researcher is face-to-face with the interviewee, he should gradually encourage them to collaborate, but first he should briefly explain the nature and objectives of the research so that the information collected becomes more consistent and reliable, as it is necessary to extract as many accounts as possible that are consistent with reality.

INTERVIEW script

1) Do you have an IDER stove in your home? Since when?

2) By what name do you know this stove?

3) And why do you think it's called that?

4) What kind of stove do you still have at home? Which one(s) do you currently use (apart from the ecological one)?

5) Do you like the eco stove? Why do you like it?

6) And what don't you like about the eco stove?

7) What could the eco stove improve?

9) Who collects the firewood in the woods? Have there been any complaints?

10) When did you first hear about the eco stove?

11) Have you attended any courses/meetings on the stove?

12) What topics were covered during the course/meeting?

13) Has the eco-stove brought any changes to your life? What changes?

Did you feel any difference in your health and the health of your family when you used the ecological fire?

14) In your opinion, is using this stove a way of helping nature? Why?

5. RESULTS AND DISCUSSION

5.1 General aspects of using ecological fires

The women chosen to benefit from the fires took part in courses / meetings. After these meetings, they were asked what advantages the technicians had for them. The answers were as follows: that the stove was good, that it would be very useful; that they could use other things to make the fire, such as cardboard, clay, coconut shavings, pieces of wood; that you could make anything with this stove, such as cake, eggs, tapioca, meat; that it cooked faster; that they didn't have to keep blowing on it to get the fire going; that it didn't "stain" the pots, the walls or the roof tiles, because the smoke didn't stay inside the house; that it was good for the family's health, because it would reduce respiratory diseases and allergies; that it would save money, because it would reduce the use of firewood; they also warned people to be careful when opening the stove, because the large amount of heat could cause people to burn themselves.

Of the interviewees who took part in the meeting, only two mentioned issues relating to the environmental theme dealt with at the meetings: "the technician said that this stove didn't pollute the environment or that it polluted less and that this would improve the environment and the housewife's life". Here it is necessary to emphasize that we seek to understand the environment in its complexity, trying to understand it in a balanced relationship between society and nature, in a broader and more complete conception of the environment and all the relationships that are established around us.

At the end of the meeting, the technician proposed that the women take the test. He asked them to confirm who really wanted to receive the stove and, if they didn't, to give it to someone else. Faced with this proposal, some of the women said that they had actually received it to test it out, to get to know it, to see if it was any good. Considering that this was the perspective of the beneficiaries, a study like this is essential to find out what their analyses were after using this technology.

In order to understand women's views on the environment, thermal comfort and improvements in quality of life brought about by ecological fires, which the beneficiaries learned about through the course mentioned above, 25 interviews/conversations were held with the beneficiaries of these fires in the community of Garapa - 1, located in the municipality of Acarape, Ceará. In these interviews, the questionnaire respondents shared their experiences and had the opportunity to explain in detail their experience of using ecological fires.

The first question was about the year in which this technology was installed. The beneficiaries of the ecological fires found it difficult to determine the year the technology was installed. Of all the respondents, only 8% could easily remember the year, while 60% could not remember the period in which the fires were installed (Figure 5). On the other hand, 32% of the respondents remembered the year the fireplaces were installed after relating the period to some moment in their lives (marriage, children, departure or death of a family member, etc).

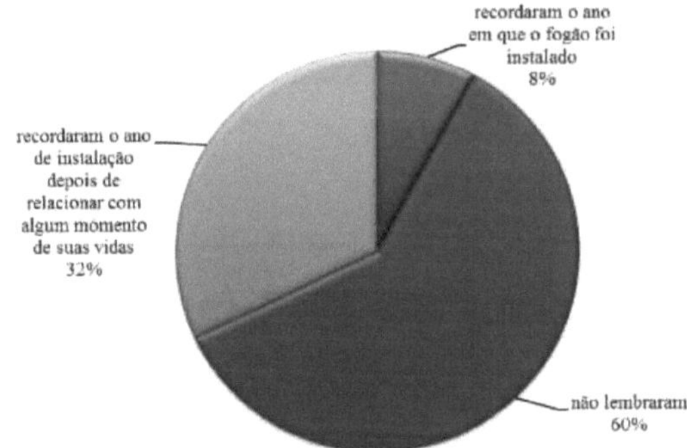

Figura 5. Women's memories of the year the ecological fires were set up in the Garapa - I community in Acarape - CE.

Source: author

In order to find out the women's basic level of knowledge about the stove project and their relationship with this new technology, they were asked by what name they knew the stove they had received. The answers were surprising: 60% of the respondents couldn't remember the name of the stove given to them by the project and the institution responsible for it, 10% could remember the name of the stove and 30% only knew it by the nickname, fogao precoce, given to the ecological stove by the community (Figure 6). According to the people who live in the community, this term is due to the fact that the stove cooks food more quickly. The renaming and reconceptualization of the stove by the community can provide interesting reflections on how people reinterpret and try to relate what comes from outside to their reality, using terms that are closer to their daily lives.

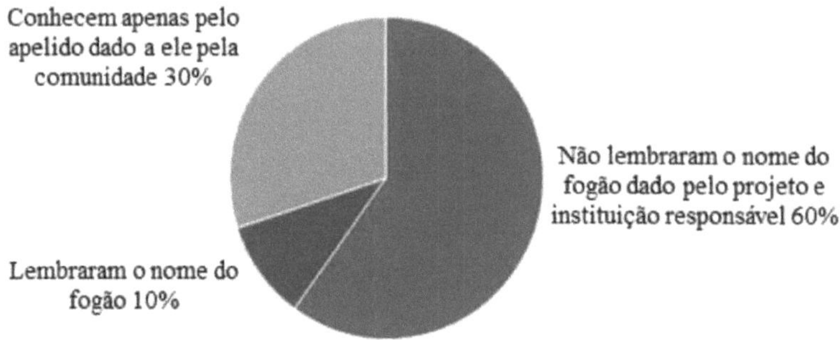

Figura 6. Women's knowledge of the ecological fire project in the Garapa - I community in Acarape - CE.

Source: author

When we tried to find out when the women first heard about the ecological stove, we wanted to find their initial memory of this relationship, of the process of coming into contact with this technology. All the beneficiaries said that they heard about the stove from the local leader (Mrs. Macleyde), when she informed them about the project and invited them to the meeting.

According to the beneficiaries, it was at the meeting with the technician from the Institute for Sustainable Development and Renewable Energies (IDER) that they saw and learned about the structure of the stove and how it worked. Before the technician came, they hadn't heard of this technology, which is understandable, given that the project was in its infancy and that the community was one of the first to benefit from this technology in the state of Cearà. Some of the women said that after the fires were installed, they started to see TV reports about them.

The local leader, who brought the information to the rest of the community, says that before the

Acarape union representative told her about the fires, she had never heard of them either. She also recalls that it was her stove that was the first to be built and that it was at her house that all the masons came to learn how the fires would be built.

5.2 Advantages of using ecological fires

With the aim of detecting women's level of satisfaction with ecological stoves, the following questions were asked: Do you like the ecological stove? Why do you like it? And what don't you like about the ecological stove? They were also asked to make suggestions as to how the stove could be changed to improve its use and possible efficiency. In these questions we tried to list all the aspects explained about the advantages of ecological stoves, what women like most about these stoves, showing the percentage of women who considered each aspect.

Among some of the advantages presented by the beneficiaries of the stoves, the fact that the stove cooks faster than a traditional wood stove was mentioned by 72% of the interviewees, who pointed out that once the fire is hot it stays that way for a long time. The issue of smoke was reported by 40%, who said they liked the stove only at the beginning, when the smoke was blown out of the house through the chimney, but that the chimney clogged easily and then the smoke returned into the kitchen. They also mentioned how important it was not to have this smoke in the house for their health and that of their children. The fact that the stove doesn't leave dirt on the pans or the walls of the kitchen, keeping it cleaner and giving women less work, was mentioned by 44% of the women. Having less work is undoubtedly a quest for these women. Only 16% of the women cited the reduction in firewood consumption as an advantage. This low percentage, however, shows us that what is presented as a fundamental aspect of the stove, giving it the name of eco-efficient, has not been identified by these women.

Savings in the consumption of butane gas were mentioned by 12% of the beneficiaries, since they prepare food that needs more time to cook, such as beans, on the eco stove, and they end up using the eco stove more often when they don't have the money to buy gas.

It can be seen that the most important issues considered by the women interviewed to be positive features of the stove are those that help to reduce their workload, making food production faster or making it easier to clean the kitchen space or utensils. Another aspect that is of great importance to them concerns the health of their children, another concern that is related to their relationship with the work of caring for the family, which generally falls to them.

With regard to chimneys and their smoke, it can be seen that the women relate pollution to the subject: the fact that they said that the stove polluted the environment less because the smoke "went upwards" indicates that their idea of the environment is strongly related to the place where they live directly, with which they relate on a daily basis, and there is no broader understanding of what environmental pollution in the world, in the atmosphere, might be.

The impression is that without more information and more in-depth debates on the main topic, the beneficiaries continued with the perception that the world is what I see, what I know. This suggests that there was no ongoing process of environmental education to help change the way these women looked at the environment, and to help broaden their perceptions, including having more elements to differentiate the ecological stove from the traditional stove. For Jacobi (2003), environmental education is increasingly taking on a transformative role, in which the co-responsibility of individuals becomes an essential objective in order to promote a new type of development - "sustainable development", which must be a process of permanent learning, valuing the various forms of knowledge in a holistic perspective of action, which relates man, nature and the universe, bearing in mind that natural resources are exhaustible and that the main person responsible for their degradation is the human being.

During the conversations, we observed that the relationship they have with the logic of efficiency is much more related to the ease and agility that this technology can bring to their lives and the reduction in the amount of work for which they are responsible. The reference to efficiency with an environmental logic does appear occasionally. This could indicate that if socio-environmental projects were concerned with understanding how unequal gender relations take place, they would know how to respond and bring improvements to people's lives both from the place and the work they do and could propose more structural changes in order to build relationships of greater respect both between people and between them and the environment.

It is worrying when the discourse of eco-efficiency or ecological efficiency prioritizes the economic aspect in an attempt to relate it to the discourse of seeking environmental protection. Sometimes, the human side of people's desires and real needs are not seen as the starting point for projects or social technologies to be recognized as transformative alternatives, which are understood and defended by the beneficiaries as something that has changed their quality of life for the better.

5.3 Disadvantages of using ecological fires

To identify the disadvantages of the stove, the women were asked what they didn't like about the stove and what bothered them. The percentage of times an aspect was mentioned is also shown here. Of the women interviewed, 56% complained that the chimney pipe is narrow, clogs too quickly and that the smoke ends up coming back into the house, polluting the environment and dirtying the pots and pans. The issue of the stove heating up was mentioned by 44% of the beneficiaries, who said that what bothers them is the fact that the stove gets too hot, making it difficult to stay near it for long. Some cited cases of family members getting burned or health problems that they related to the excess heat. As for having to cut the firewood smaller, chopping small bundles, 32% of the women cited this as a difficulty. In this respect, we can see that it is largely the husbands who complain that this technology has made more work for them. The difficulty of cleaning the stove was cited by 18% of the women interviewed, who explained that the space to remove the ash is very tight, making it harder to unblock (Figure 7).

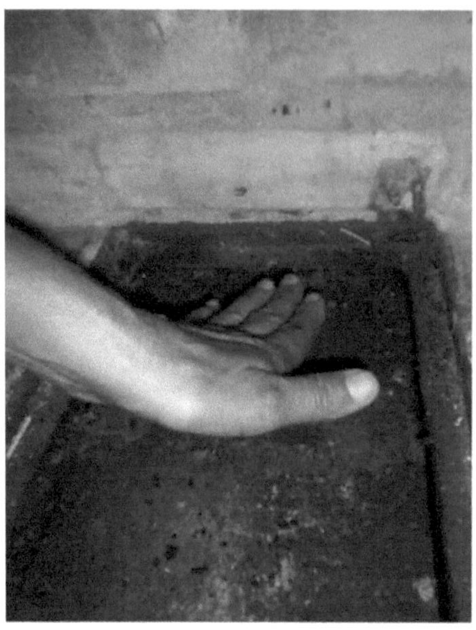

Figura 7. Tray where the ashes from ecological fires are collected in a house in the Garapa - 1 community in Acarape - CE.

Source: author

The longer time it takes to prepare the food was cited by 8% of the interviewees as a disadvantage: "You have to start lighting the fire earlier, because it takes so long to heat up", and they also considered it very easy to burn themselves on it. The small size of the griddle was mentioned by only one of the interviewees.

It can be seen that some of the complaints are visibly the result of inappropriate use or use that differs from the way the team was instructed to use it at the time of the meeting with the community. This can be seen when the negative points are related to the dirt that the stove leaves in the kitchen, for example. This is because they use the stove in the same way as the traditional stove: many of them don't cut the wood into smaller pieces, preferring to use large pieces of wood and keeping the little door, which should be closed, open, which keeps the smoke circulating in the kitchen and the resulting dirt. Some of the users even ripped the little door off because, according to them, it was more work to cut the wood into smaller pieces (a complaint usually made by the men who do this work).

The users raise questions related to the technical elements or the way the fires were built, which they believe compromises their use. One example is the fact that they consider the chimney pipe to be too narrow, which according to them means that it clogs up too quickly and soot gets back into the homes. The technical eye would probably say that it's the women who aren't cleaning with the proper frequency, that they aren't tapping the chimney to loosen the soot and make it easier to clean. But, in the eyes of the women, this makes it less efficient, because it requires a greater amount of work.

They also complain about the space to remove the soot that falls, which is too small, making it difficult to clean, and that the space to place the firewood is large and could be smaller and require less firewood. It's true that others would prefer the space to be larger, to fit more wood and no longer need to cut such small pieces. Some of the questions raised here don't really have a technical perception, but they can make a contribution, since the aim is to give visibility to empirical knowledge, based on everyday use, experiences and experiments.

5.4 Use of ecological fires and environmental aspects

In the first interviews, the women found it difficult to understand the concept of the environment. They were asked to talk about how they thought the stove contributed in some way to caring for and helping to preserve nature. It was interesting to see how what seems like such a popular concept is still not understood by some people and to think about how concepts, terms created by academia, by scholars, take a long time to become understood and actually incorporated by "common sense".

It was difficult to record the percentages for this item, because the women brought elements that they believed contributed and elements that they felt did not improve environmental conditions. The answers were also very different. Of those interviewed, 32% said they couldn't explain the relationship between the stove and nature and another 20% thought that the stove didn't contribute to the environment. While 48% of respondents believed that stoves contribute to the environment. The arguments cited in relation to the benefits and harms they identified of the stove for the environment will be presented below.

With regard to the contributions that the stove and its use have made to nature, the two aspects mentioned by the most women were the reduction in pollution and smoke and the saving of firewood. Some said that there was very little smoke coming out of the chimney and that for this reason the stove didn't pollute much. One added that with this amount "you can't contaminate the trees". They also commented that "its smoke isn't very polluting" and that it pollutes less "because it doesn't just burn wood, it can also burn the cobs, the corn cobs, so it cleans up the environment". Another interesting comment related the reduction in pollution to the fact that the smoke, which used to come out scattered, now comes out all together (in the chimney). We also heard that "it helps, because the smoke comes out of the house, it disappears in the middle of the world, it evaporates" and that "it doesn't harm because the smoke goes upwards". Few were able to distinguish between internal and external pollution, understanding that "the only thing that has changed is inside the house, no more smoke inside the house".

It can be seen that the idea of environmental pollution, so often presented in different media, seems to be difficult to understand, especially for older women, who generally have a low level of education. Thinking about what pollution is; which elements pollute less and which pollute more; relating the pollution we see and feel in the environment we are in with that which is not visible, which is in the air, is indeed complicated. Younger and more educated people are explicitly more likely to understand how pollution occurs and what impact it can have on their lives.

Environmental legislation is recent and has helped us to understand the difficulties people still have in becoming concerned and adopting new daily practices to conserve the environment. As well as learning a little about the provisions of this law, it raises some questions: are we really carrying out a process of environmental education? How can we contribute, including through non-formal education, to this change in individual and collective values, skills and competences? What impact can one-off, non-continuous actions to raise awareness of environmental issues have on changing habits?

Some were able to make interesting connections between the stove and their contribution to nature, which shows how much easier it is for them to see links from their everyday actions: they said that by reducing the use of soap they were helping nature, "because it doesn't pollute the soil" and that they help to clean up the environment and prevent dengue fever by using, for example, coconut shavings on the stove. "It doesn't collect mosquitoes in the water, it doesn't collect water and it doesn't generate dengue mosquitoes."

Another aspect that the women of the stove relate to the preservation/conservation of nature is the use of firewood. The perception of saving firewood is not consensual, but some believe that it does save more firewood, linking it to a reduction in deforestation. One commented that she thinks it saves money "because the wood is small" and they can pick up the sticks that are loose on the ground. However, those who don't think the stove contributes to preserving nature said: "It doesn't help nature, because it uses more wood than the other kind. It takes more wood to heat it up, I thought it wasted more. You have to put a lot of wood in, two batches to heat it up and then put more in. It takes more wood to heat them up because more wood fits in the container intended for this purpose. Sousa 2007, reports concern about the negative effects of using the caatinga as an energy source. The woody species that proliferate in plant communities are being devastated, and their extraction is used for a variety of purposes, ranging from domestic consumption to use in pottery.

It is necessary to find out in more detail why the perception of whether or not firewood is saved is so different. For this, more specific research is suggested, in which it is possible to accurately calculate the consumption of firewood in m^3 .

Considering that this is a fundamental pillar for reflection: what kind of impact do these fires have on deforestation if the main cause of this action, based on the example of this community, is not domestic use, but commercial use? Wouldn't the ideal be to invest in alternatives that change the energy matrix? What risk do we run when we place the responsibility for the depletion of natural resources, such as water and firewood, on individuals, when it is companies and businesses that consume the most? How can we think about changes in individual attitudes if they are not accompanied by structural transformations in the current logic of consumption?

Given these women's perceptions of the environment, we can see the importance and need for a social process of environmental education which, based on Paulo Freire's references (FREIRE, 1979), states that "education is essentially an act of knowledge and awareness". This, however, is not a neutral act, it needs to be liberating, transformative, because being a political act it contributes to raising awareness of the contradictions of the human world and our power to change this reality, for example, of so much environmental destruction.

5.5 The use of ecological fires and the temperature in the room

A very relevant issue that is always mentioned by the women is the high temperature of the fireplaces, which often makes it difficult for users to get close to them and stay there. It was found that during their use in some cases, the ambient temperature varied between 33.1 °C and 45.0 °C (Figure 8) and the temperature at different points in the structure of the stoves reached up to 130 °C and that they remain hot until dusk, which ends up contributing to a certain amount of thermal discomfort, because as well as having to stay close to the stoves during food preparation, the semi-arid climate of the region contributes to thermal discomfort. Perhaps this was one of the reasons why the women rejected ecological fires so much. Only two women reported that this heat was good on the one hand, because they left the pans on the stove, which had already been turned off, and the food remained "warm" until their husbands got home from work, so they didn't need to heat up the food.

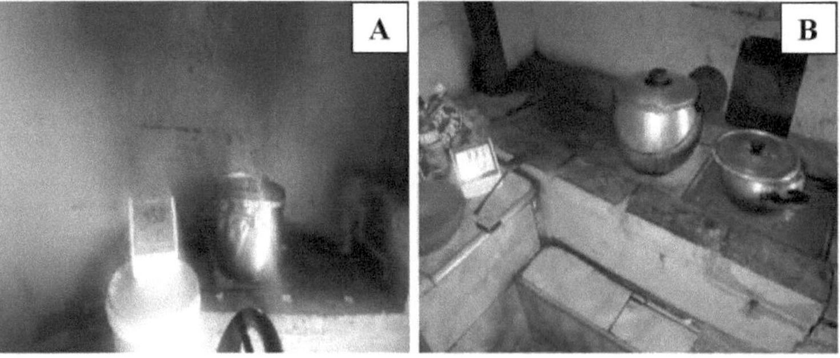

Figura 8. Thermometer marking the ambient temperature during the use of the ecological stove (A

and B) in a residence in the Garapa - I community in Acarape - CE.

Source: author

Analyzing the perception of heat in the environment when the stove is lit, whether it's a traditional wood stove, an ecological stove or a gas stove, in (Figure 9) you can see the heat variation in the place, according to the data reported by the residents of the dwellings. We can see that the temperatures are high and that it is really difficult to stay near these fires for long.

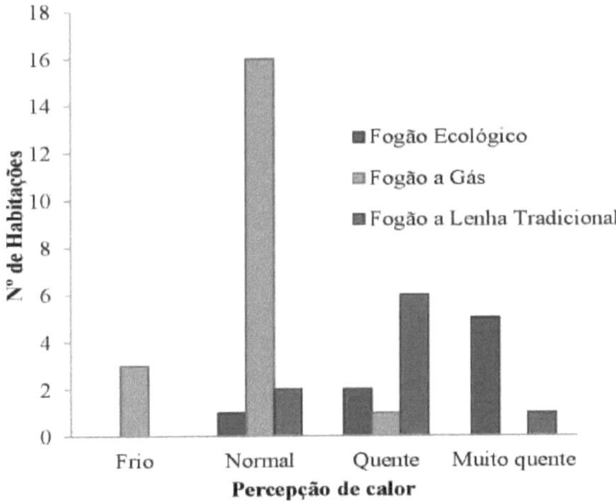

Figura 9. Perception of heat by residents of ecological stove dwellings in a residence in the Garapa - I community in Acarape - CE.

Source: author

5.6 Current situation regarding the use of ecological fires

Another important piece of information for the analysis was the current state of ecological fires, which expresses the degree of acceptance of this technology by the beneficiaries.

The majority of those interviewed, 80%, said they had dismantled the stove: 8% had completely scrapped it; while 32% had dismantled the ecological stove, but had rebuilt it in another format, removing parts of the stove that had come through the project, adapting it to the way they considered best, generally the way closest to using the traditional wood stove. 32% completely dismantled the stove because they weren't satisfied with it due to various factors, such as the high temperature, the chimney that clogs too much and the extra work involved in cutting the wood into small pieces. Around 8% of those interviewed said that they dismantled the stove and rebuilt it in another place, but keeping the same shape and original characteristics, and only 20% kept the stove the way they received it (Figure 10).

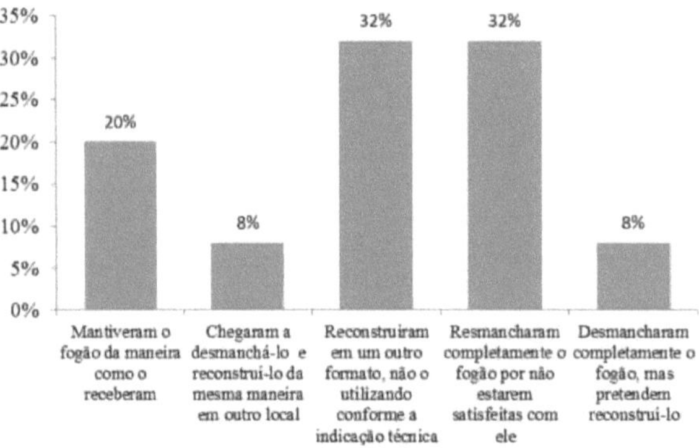

Figura 10. Current situation of ecological fires in the Garapa - I community in Acarape - CE. Source: author

Figure 11 shows some examples of the previous data, ecological fires that have been completely dismantled and/or abandoned.

Figura 11. Unused ecological fires in three different homes (A, B and C) in the Garapa - I community in Acarape - CE.

Source: author

Based on the data reported, a central issue was identified: the difficulty of changing consolidated social practices that had already become customs, loaded with their own knowledge. When questioning the beneficiaries about this difficulty, it was understood that custom is built on the constant and uniform repetition of a social practice that has worked over time.

According to Nader (2011), custom is the effective norm that aspires to validity. An example of this is how communities and people used equipment and resources that had proved efficient up to that point for cooking food, before the arrival of new technologies. As soon as someone brings a new technology, a new piece of equipment into kitchens, such as the ecological stove, as being more efficient than the wood stove traditionally used, doubts and questions arise about the new compared to the usual.

It is known that changes in practices are possible within a certain historical context of a community's social life. The same practices are used until new ones are not only introduced, but are proven to be better, more efficient. It has to be worth doing differently. And the question remains: were the women really convinced, not just through arguments but also through use, that this new technology was more efficient and better for the environment? Were they able to understand the difference between the traditional wood stove and the ecological or eco-efficient wood stove? It's also interesting to realize that, in general, the projects come to the communities with their proposals ready-made, already defined and not very open to the different relationships that people have with their spaces, their work and the environment. In order to continue to understand how these women relate to the stove in their daily lives, and how they use it, we sought to understand the degree of acceptance of the ecological stove. The data shows a worrying reality: 60% of the women interviewed do not use the stove they received, 8% use it only occasionally, while 32% use ecological stoves frequently (Figure 12).

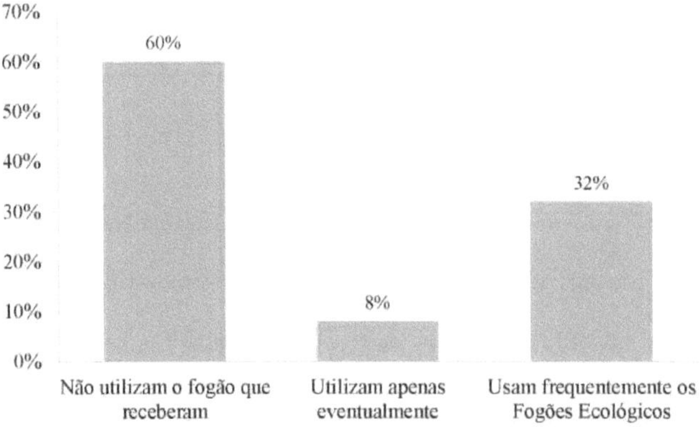

Figura 12. Level of acceptance of ecological fires in the Garapa - I community in Acarape - CE.

Source: author

With regard to the data on what type of stove the women interviewed currently use, 8% of them still use 3 types (traditional wood, ecological and gas). They said that they prefer and use the traditional stove more often, using the ecological stove occasionally, when they are going to cook for a longer period of time, or when they have a party at home. On the other hand, 32% use the ecological stove and the gas stove in parallel (Figure 13), and 60% no longer use the ecological stove they received, using only the traditional wood-burning stoves and the gas stoves. This percentage includes all the beneficiaries who have altered, dismantled or use the stove in the traditional way, even if they still maintain some of the structure of the stove received by the project.

Figura 13. Use of gas and ecological stoves in a home in the Garapa - I community in Acarape - CE.

Source: author

It's important to note that most of the women who stopped using the ecological stove only used it for a short time (between 3 months and 1 year). When the stoves started to show problems, especially with the chimney, and the smoke returned to the inside of their homes, they stopped using them. But the interesting thing is that they continued to use the traditional fires, which had the same problem with the smoke inside the house.

The abandonment of the use of the ecological stove is a fundamental issue to be analyzed. It could be related to the difficulty of changing habits, practices, reinforced by the lack of time or space to understand the difference between this stove and the traditional one, or it could be the result of a negative evaluation of the technology, probably not seeing real efficiency in its use.

5.7 Suggestions for improvement

Faced with the disadvantages presented above, the women interviewed offered some suggestions for improving the ecological stove, based on their perceptions and needs. To ensure that no comments are left out, all the proposals mentioned have been listed.

The suggestion that the technicians think of a way for the stove to heat up less, by containing and reducing the heat received by them, was said by most of them, 68%.

The proposal to increase the thickness of the chimney pipe and to have more space between the base of the chimney and the worktop to make it easier to clean and remove soot was mentioned by 24% of the women.

Of the women interviewed, 40% would like to increase the size of the griddle, taking into account the size of the family, as well as the worktop, allowing more space for handling pans and food. There was also a suggestion from 24% of the women that it would be interesting if the stove had an oven. Some of them, 12 %, would like the base of the stove to be higher or to take into account the average height of the family or the people who would use the stove the most.

There was no consensus on just one proposal: of the women interviewed, 64% believe that the space for placing firewood should be larger, so that they can place larger pieces of firewood and don't have to cut the wood into smaller pieces, while 36% believe that the space should be smaller, so that they need less firewood.

These observations and/or suggestions made by the beneficiaries of the ecological fires serve to reflect on the relationship between scientific knowledge and empirical knowledge. Generally, in the execution and implementation of government policies and programs, one notices how much theoretical, academic knowledge, which is considered to be correct, prevails over experience and popular knowledge. Unfortunately, there is no dialogue between these two types of knowledge: the project technicians have already predefined what they need to do and have little time to listen to the people, their needs, their perceptions and their common sense. According to Babini, 1957, popular knowledge, *lato sensu,* "is acquired in direct dealings with things and humans, it is the knowledge that fills our daily lives and that we possess without having searched or studied, without applying a method and without having reflected on something".

This is a very important aspect to consider for any governmental or non-governmental program or project: buildings like these need to take into account people's desires and daily lives, especially those who will use them frequently. It is also necessary to consider not only technical factors and financial limits, but also the way people relate to their spaces and to each other. Failure to consider these issues can lead to rework or wasted resources, which could be seen in the rebuilding of fireplaces elsewhere or even in some that have been dismantled.

6. CONCLUSIONS

Thus, the analysis of the general data from the survey conducted with the users of the ecological fires revealed that, despite the low-cost implementation and maintenance of the ecological fire models in the Garapa I community, both the traditional wood-burning model and the gas stove are still used on a large scale in the community. The beneficial effects on health have not been properly perceived with ecological fires, as the smoke remains inside the houses. According to the reports of the beneficiaries of these fires, at first the smoke was expelled to the outside of the homes through the chimney, but over time the accumulation of soot in these chimneys caused them to clog and consequently the smoke to return to the inside of the homes.

Thus, the analysis of the general data from the survey carried out with the users of ecological stoves showed that, although the objectives of these stoves are to reduce the consumption of firewood and eliminate smoke inside homes, it was found that these objectives are not being achieved, since most of the interviewees say they feel no difference in their health when using the ecological stove, the use of this technology negatively alters the health of the interviewees, especially breathing and the high temperature of the environment.

With regard to the temperature of the fires, the women's biggest complaint was that during their use, the ambient temperature reached 45°, while the temperature of the fire structure reached 130°. And the fire remains hot for a long time (around 6 hours) once it has been extinguished.

The importance of disseminating improved wood-burning stove technologies in rural areas is undeniable, but first it is necessary to prove that these technologies will actually meet their objectives, otherwise public money may be wasted. In addition, it is of fundamental importance that, together with programs to disseminate wood-burning stoves, there is environmental education and preservation work, especially in partnership with schools and society.

It was clear that the difficulty in changing habits and practices identified in the research may have a direct relationship with the little time given to a fundamental process: raising awareness and training about the difference between ecological stoves and traditional stoves, not just for future users of the stove, but for the community as a whole. The benefits that a change in habits could bring to people's quality of life in the medium and long term.

A more continuous process of environmental education with women, men, children and the elderly in the community could probably make it possible to use the eco-efficient stove more consciously and generate other community actions to preserve their forests against deforestation and burning.

As a suggestion for future work, comparing the consumption of firewood in m3 between the two fires (traditional and ecological) would be very important to evaluate the efficiency of these fires.

7. BIBLIOGRAPHICAL REFERENCES

ALCÓCER, J. C. A.; DUARTE, J. B. F.; M. J. CAJAZEIRAS; M. L. M. de Oliveira ; R. G. Duarte ; ROCHA, Y. M. G. ; PONTES, B. C. M. ; J. DUARTE ; I. HOLANDA ; QUEIROZ, D. M. B. ; RAMOS, K. M. ; J. O. DIOGO ; G. N. DANTAS. **Producing Biogas from Fruit Waste to Generate Electricity**. SODEBRAS Magazine, v. 9, p. 113-116, 2014.

ALCÓCER, J. C. A.; DA COSTA, J. M. F.; RAMOS, K. M.; DUARTE, A.; MOREIRA, K.; COAQUIRA C. A. C.; GUIMARAES, A. P.; DUARTE, J. B. F. **Treatment of Domestic Sewage from Rural Regions with Evapotranspiration Tanks**. SODEBRAS Magazine, v. 10, p. 22-25, 2015.

ANDRADE, Daniel Fonseca de; LUCA, Andréa Quirino de; SORRENTINO, Marcos. **Dialogue in environmental education public policy processes in Brazil**. Rev. Educ. Soc., Campinas, v. 33, n. 119, p. 613-630 - Apr/Jun, 2012.

BABINI, J,. **Knowledge.** Buenos Aires: Nueva Visón, 1957.

BACCHI, M. R. P. **Generating clean and renewable biomass energy**. 2006.

BEN/MME. **National Energy Balance: Base year 2010** (2011) Rio de Janeiro: Empresa de Pesquisa Energética.

BERMANN. C. **Environmental Crisis and Renewable Energies**. Ciência e Cultura, V. 60, n°. 3. Sao Paulo, 2008.

BITTENCOURT, H. V. H. **A matriz energètica no desenvolvimento sustentâvel de pequenas propriedades rurais**. TCC - UFSC - CCA. 2005.

BORGES, T. P. de F. **Clean-burning wood stove**. State University of Campinas - SP, 1994.

BRITO, R.C. CORREIA, A.C. CAVALCANTI, F.H.B.B. SILVA,J.C. ARAJO FILHO & A.P. LEITE . 1997. **Agro-ecological zoning of the Northeast: Diagnosis of the natural and agro-socio-economic framework**. EMBRAPA CPATSA, Petrolina, PE.

BRUNDTLAND REPORT. **Our Common Future**. New York, Oxford University Press, 1987.

CAMPELLO, F.B.; GARIGLIO, M.A.; SILVA, J.A.; LEAL, A.M.A. **Diagnòstico Aorestal da regiao Nordeste**. Brasilia: IBAMA/PNUD, 1999.

CARVALHO, R.L.T. SILVA, A. C. SANTOS, P.G.L. and TARELHO, A. L. C. **Comprehensive study of environmental comfort in rural housing in Ceará.** Revista GEONORTE, Special Issue 2, V.2, N.5, Manaus, 2012.

DRUMOND, M.A, et al. **Evaluation and identification of priority actions for the conservation, sustainable use and sharing of benefits from the biodiversity of the Caatinga biome.** Petrolina: Document for discussion in the WG Strategies for Sustainable Use, 2000.

FREIRE, Paulo Reglus Neves. **Education and Change.** Sao Paulo: Paz e Terra, 1979.

INSAURRALDE, P. A. B., ALCÓCER, J. C. A., NUNES, A. B. S., CORREIA, E. P., NASCIMENTO, E. R. M., BATISTA, M. K. S. **Fogo Ecológico: Uma Alternativa Eficiente para Substituição dos incêndios** A **lenha convencionais**. Environmental Education in Action. , v.52, p.4,2015.

IDER - Institute for Sustainable Development and Renewable Energies. **Eco-Efficient Fires**, 2012. <http://www.ider.com.br>. Accessed on: August 23, 2014.

IBGE - Brazilian Institute of Geography and Statistics. Profile of Cities. Available at: <http://www.ibge.com.br>. Accessed on: January 23, 2016.

JACOBI, Pedro. **Environmental education, citizenship and sustainability**. Cadernos de Pesquisa, n° 18, 2003.

KAMIMURA, Arlindo; BURANI, Geraldo F. 2010 **Energia, Economia, Rotas Tecnológicas.** *Selected texts*, Free electronic edition. Full text at <www.eumed.net/libros/2010e/827/>Accessed on: December 15, 2015.

KATWAL, R.P.S. and P.L. SONI. 2003. Biofuels: an opportunity for so-cioeconomic development and cleaner environment. Indian For. 129, 939-949.

LEAL, I.R., M. Tabarelli & J.M.C. Silva. **Ecology and conservation of the Caatinga**. Recife: University Press, 2003.

MACHADO, Marilia Novais da Mata. **Psychosocial research and intervention.** Proceedings of the 12th National Meeting of Abrapso - Brazilian Association of Social Psychology. Porto Alegre: Puc/RS, Abrapso. CD Rom, 2003

MAIA, G. et al. **Implementation of a dissemination strategy for efficient cook stoves in northeast Brazil**, Institute for Sustainable Development and Renewable Energies, 2008.

MARTINS, G.; BARROS, I. F. R.; LIMA, M. D. Social design vs. new technologies. Proceedings of the International Workshop on Technological Renewal, Curitiba, PR, 1992.

MMA - Ministry of **the** Environment, 2013 - **The efficiency of ecological fires** <http://www.mma.gov.br/informma/item/9489-a-eficiència-dos-fogòes-ecológicos>Accessed: 15 Jan. 2016.

MORAES, M. M.; MARTINS, G.; TRIGOSO, F. B. M. 2007. **The use of wood-burning stoves and the semi-arid region of Piauí: A case study**. Federal University of ABC. 2007.

NADER, Paulo. **Introduction to the Study of Law**. 33 ed. Rio de Janeiro: Forense, 2011.

REIS, L. B. dos and SILVEIRA, S. (eds.). **Electric Energy for Sustainable Development**. Sao Paulo: Edusp, 2000.

SANGA, G. A. **Impact assessment of clean technologies and substitution of cooking fuels in urban households in Tanzania**. UEC. 2004.

SATTLER, Miguel Aloysio. **Sustainable buildings and communities: activities under development at NORIE/UFRGS**. V Seminar on Transfer and Training for Social Interest Housing, 2003.

SILVA, Roberto Marinho Alves. **Between Combating Drought and Coexistence with the Semi-Arid: Paradigmatic Transitions and Sustainability of Development** (Doctoral Thesis). Brasilia:

UNB, 2006, 298p.

SIMON, L. G.et al. **Win-win scenarios at the climate-development interface: Challenges and opportunities for stove replacement programs through carbon finance**, Global e nvironmental Change, university of Colorado Denver, USA, 2011.

SOUSA, R. F. de. **Agricultural land and the desertification process in municipalities in the semi-arid region of Paraíba**. 2007. 180f. Thesis (Doctorate in Agricultural Engineering) - Center for Technology and Natural Resources, Federal University of Campina Grande, Campina Grande.

SOUZA, R. C. R. PEREIRA, G. A. FRANÇA, B. S. MARTINS, G. **Improvement and Dissemination of Clean-Burning Wood Stoves in the State of Amazonas**. In: Proceedings of the 3rd Meeting on Energy in the Rural Environment. Campinas - SP, 2003.

PARIKKA, M. **Global biomass fuel resources**. Biomass and Bioenergy, v.27, n.6, p.613620, 2004.

PIMENTEL, Alamo. **O Elogio da Convivência e suas Pedagogias Subterrâneas no Semiàrido Brasileiro**. (Doctoral thesis). Porto Alegre: UFRGS, 2002. 341f.

UHLIG, A. **Firewood and charcoal in Brazil: supply-demand balance and methods for estimating consumption**. PhD thesis - USP. 2008.

UNIVERSITY OF SÂO PAULO-USP. Database of Biomass in Brazil. **Firewood in Brazil**, 2013. Historical Series. Available at:

<http://infoener.iee.usp.br/scripts/biomassa/br_lenha.asp>. Accessed: March 28, 2016.

VESENTINI, José William. **The New World Order**. Sao Paulo: Atica, 2000.

VIEIRA, V. P. P. B. **Desafios da gestão integrada de recursos hidricos no semiàrido**. Revista Brasileira de Recursos Hidricos. V. 8, p.7-17, 2003.

VITOUSEK, P., J. Aber, R. Howarth, G. Likens, P. Matson, D. Schindler,W. Schlesinger, and D. Tilman. 1997. **Human alteration of the global nitrogen cycle: causes and consequences.** [COMPLETE].

WINTROCK INTERNATIONAL BRAZIL. International Meeting on Odorless Air Pollution, Fuel Efficient Stoves and Sustainable Development. Salvador, 2007. Available at <http://pdf.usaid.gov/pdf_docs/PNAD1781.pdf>.

ZANELLA, Maria Nilvane. **The UN perspective on minors, offenders, delinquents and adolescents in conflict with the law: social education policies**. Master's dissertation in Education - Maringà State University, 2014.

Printed by Books on Demand GmbH, Norderstedt / Germany